Index

What is a Planet?

This seemingly simple question doesn't have a simple answer. Everyone knows that Earth, Mars and Jupiter are planets. But both Pluto and Ceres were once considered planets until new discoveries triggered scientific debate about how to best describe them—a vigorous debate that continues to this day. The most recent definition of a planet was adopted by the International Astronomical Union in 2006. It says a planet must do three things:

1. It must orbit a star (in our cosmic neighborhood, the Sun).
2. It must be big enough to have enough gravity to force it into a spherical shape.
3. It must be big enough that its gravity cleared away any other objects of a similar size near its orbit around the Sun.

Planets

There are more planets than stars in our galaxy. The current count orbiting our star: eight.

The inner, rocky planets are Mercury, Venus, Earth and Mars. The outer planets are gas giants Jupiter and Saturn and ice giants Uranus and Neptune.

Beyond Neptune, a newer class of smaller worlds called dwarf planets reign, including longtime favorite Pluto. Thousands more planets have been discovered beyond our solar system. Scientists call them exoplanets (exo means "from outside.")

Dwarf Planets

They key difference between a planet and a dwarf planet is the kinds of objects that share its orbit around the Sun. Pluto, for example, has not cleared its orbit of similar objects while Earth or Jupiter have no similarly-sized worlds on the same path around the Sun. Like planets, dwarf planets are generally round (Haumea looks like an overinflatred football) and orbit the Sun.

There are likely thousands of dwarf planets waiting to be discovered beyond Neptune. The five best known dwarf planets are: Ceres, Pluto, Makemake, Haumea and Eris. Except for Ceres, which lies in the main asteroid belt, these small worlds are located in the Kuiper Belt. They're considered dwarfs because they are massive, round, and orbit the Sun, but haven't cleared their orbital path.

What is a Moon?

Moons – also known as natural satellites – orbit planets and asteroids in our solar system. Earth has one moon, and there are more than 200 moons in our solar system. Most of the major planets – all except Mercury and Venus – have moons. Pluto and some other dwarf planets, as well as many asteroids, also have small moons. Saturn and Jupiter have the most moons, with dozens orbiting each of the two giant planets.

Moons come in many shapes, sizes, and types. A few have atmospheres and even hidden oceans beneath their surfaces. Most planetary moons probably formed from the discs of gas and dust circulating around planets in the early solar system, though some are "captured" objects that formed elsewhere and fell into orbit around larger worlds.

Asteroids, Comets & Meteors

At the very beginning of our solar system, before there was an Earth, Jupiter or Pluto, a massive swirling cloud of dust and gas circled the young Sun. The dust particles in this disk collided with each other and formed into larger bits of rock. This process continued until they reached the size of boulders. Eventually this process of accretion formed the planets of our solar system.

Billions of small space rocks never evolved. Amazingly, many of these mysterious worlds have been altered very little in the 4.6 billion years since they first formed. Their relatively pristine state makes the comets, asteroids and some meteors wonderful storytellers with much to share about what conditions were like in the early solar system. They can reveal secrets about our origins, chronicling the processes and events that led to the birth of our world. They might offer clues about where the water and raw materials that made life possible on Earth came from.

Current Counts:

Known Asteroids:	10,68,721
Known Comets:	3,716

Our Solar System

Image Credit :- NASA

Our Solar System

Our planetary system is located in an outer spiral arm of the Milky Way galaxy.

Our solar system consists of our star, the Sun, and everything bound to it by gravity – the planets Mercury, Venus, Earth, Mars, Jupiter, Saturn, Uranus and Neptune, dwarf planets such as Pluto, dozens of moons and millions of asteroids, comets and meteoroids. Beyond our own solar system, we have discovered thousands of planetary systems orbiting other stars in the Milky Way.

Why Is It Called The "Solar" System?

There are many planetary systems like ours in the universe, with planets orbiting a host star. Our planetary system is named the "solar" system because our Sun is named Sol, after the Latin word for Sun, "solis," and anything related to the Sun we call "solar."

The Sun

Image Credit :- NASA

The Sun

The Sun—the heart of our solar system—is a yellow dwarf star, a hot ball of glowing gases.

Its gravity holds the solar system together, keeping everything from the biggest planets to the smallest particles of debris in its orbit. Electric currents in the Sun generate a magnetic field that is carried out through the solar system by the solar wind—a stream of electrically charged gas blowing outward from the Sun in all directions.

The connection and interactions between the Sun and Earth drive the seasons, ocean currents, weather, climate, radiation belts and aurorae. Though it is special to us, there are billions of stars like our Sun scattered across the Milky Way galaxy.

Sun Facts

•The Sun is the largest object within our solar system, comprising 99.8% of the system's mass.

•The Sun is located at the center of our solar system, and Earth orbits 93 million miles away from it.

•Though massive, the Sun still isn't as large as other types of stars. It's classified as a yellow dwarf star.

•The Sun's magnetic field spreads throughout the solar system via the solar wind.

Mercury

Image Credit :- NASA

Mercury

The smallest planet in our solar system and nearest to the Sun, Mercury is only slightly larger than Earth's Moon.

From the surface of Mercury, the Sun would appear more than three times as large as it does when viewed from Earth, and the sunlight would be as much as seven times brighter. Despite its proximity to the Sun, Mercury is not the hottest planet in our solar system – that title belongs to nearby Venus, thanks to its dense atmosphere.

Did You Know?

Because of Mercury's elliptical–egg-shaped–orbit and sluggish rotation, the morning Sun appears to rise briefly, set and rise again from some parts of the planet's surface. The same thing happens in reverse at sunset.

Venus

Venus

Similar in size and structure to Earth, Venus has been called Earth's twin. These are not identical twins, however – there are radical differences between the two worlds.

Venus has a thick, toxic atmosphere filled with carbon dioxide and it's perpetually shrouded in thick, yellowish clouds of mostly sulfuric acid that trap heat, causing a runaway greenhouse effect. It's the hottest planet in our solar system, even though Mercury is closer to the Sun. Venus has crushing air pressure at its surface – more than 90 times that of Earth – similar to the pressure you'd encounter a mile below the ocean on Earth.

Did You Know?

The Soviet Union's Venera 13 survived the intense heat and crushing pressure of Venus' surface for more than two hours. Engineers from several nations are currently studying methods to extend the life of robotic spacecraft in the extreme environment.

Earth

Image Credit :- NASA

Earth

Our home planet is the third planet from the Sun, and the only place we know of so far that's inhabited by living things.

While Earth is only the fifth largest planet in the solar system, it is the only world in our solar system with liquid water on the surface. Just slightly larger than nearby Venus, Earth is the biggest of the four planets closest to the Sun, all of which are made of rock and metal.

The name Earth is at least 1,000 years old. All of the planets, except for Earth, were named after Greek and Roman gods and goddesses. However, the name Earth is a Germanic word, which simply means "the ground."

Kid-Friendly Earth

Our home planet Earth is a rocky, terrestrial planet. It has a solid and active surface with mountains, valleys, canyons, plains and so much more. Earth is special because it is an ocean planet. Water covers 70 percent of Earth's surface.

Earth's atmosphere is made mostly of nitrogen and has plenty of oxygen for us to breathe. The atmosphere also protects us from incoming meteoroids, most of which break up before they can hit the surface.

The Moon

Near Side of the Moon

Far Side of the Moon

The Moon

Earth's Moon is the only place beyond Earth where humans have set foot.

The brightest and largest object in our night sky, the Moon makes Earth a more livable planet by moderating our home planet's wobble on its axis, leading to a relatively stable climate. It also causes tides, creating a rhythm that has guided humans for thousands of years. The Moon was likely formed after a Mars-sized body collided with Earth.

Earth's Moon is the fifth largest of the 190+ moons orbiting planets in our solar system.

Earth's only natural satellite is simply called "the Moon" because people didn't know other moons existed until Galileo Galilei discovered four moons orbiting Jupiter in 1610.

Facts About The Moon

- The Moon is Earth's only natural satellite and the fifth largest moon in the solar system.
- The Moon's presence helps stabilize our planet's wobble, which helps stabilize our climate.
- The Moon's distance from Earth is about 240,000 miles (385,000km).
- The Moon has a very thin atmosphere called an exosphere.
- The Moon's entire surface is cratered and pitted from impacts.

Lunar Phases and Eclipses

Image Credit :- NASA

Lunar Phases and Eclipses

What are lunar phases and eclipses?

The moonlight we see on Earth is sunlight reflected off the Moon's grayish-white surface. The amount of Moon we see changes over the month — **lunar phases** — because the Moon orbits Earth and Earth orbits the Sun. Everything is moving.

During a **lunar eclipse**, Earth comes between the Sun and the Moon, blocking the sunlight falling on the Moon. Earth's shadow covers all or part of the lunar surface.

What Are Lunar Phases?

Our Moon doesn't shine, it reflects. Just like daytime here on Earth, sunlight illuminates the Moon. We just can't always see it. When sunlight hits off the Moon's far side — the side we can't see without from Earth the aid of a spacecraft — it is called a **new Moon**. When sunlight reflects off the near side, we call it a **full Moon**. The rest of the month we see parts of the daytime side of the Moon, or phases. These eight phases are, in order, **new Moon**, **waxing crescent**, **first quarter**, **waxing gibbous**, **full Moon**, **waning gibbous**, **third quarter** and **waning crescent**. The cycle repeats once a month (every 29.5 days).

What is a Lunar Eclipse?

During a lunar eclipse, Earth comes between the Sun and the Moon, blocking the sunlight falling on the Moon. There are two kinds of lunar eclipses:

• A **total lunar eclipse** occurs when the Moon and Sun are on opposite sides of Earth.

• A **partial lunar eclipse** happens when only part of Earth's shadow covers the Moon.

During some stages of a lunar eclipse, the Moon can appear reddish. This is because the only remaining sunlight reaching the Moon at that point is from around the edges of the Earth, as seen from the Moon's surface. From there, an observer during an eclipse would see all Earth's sunrises and sunsets at once.

Mars

Mars

Mars is the fourth planet from the Sun – a dusty, cold, desert world with a very thin atmosphere. Mars is also a dynamic planet with seasons, polar ice caps, canyons, extinct volcanoes, and evidence that it was even more active in the past.

Mars is one of the most explored bodies in our solar system, and it's the only planet where we've sent rovers to roam the alien landscape. Two NASA rovers and one lander are currently exploring the surface of Mars (and a Chinese lander is set to land later this year). An international fleet of eight orbiters are studying the Red Planet from above.

These robotic explorers have found lots of evidence that Mars was much wetter and warmer, with a thicker atmosphere, billions of years ago.

Did You Know?

Mars has two moons named Phobos and Deimos.

Jupiter

Image Credit :- NASA

Jupiter

Jupiter has a long history surprising scientists—all the way back to 1610 when Galileo Galilei found the first moons beyond Earth. That discovery changed the way we see the universe.

Fifth in line from the Sun, Jupiter is, by far, the largest planet in the solar system – more than twice as massive as all the other planets combined.

Jupiter's familiar stripes and swirls are actually cold, windy clouds of ammonia and water, floating in an atmosphere of hydrogen and helium. Jupiter's iconic Great Red Spot is a giant storm bigger than Earth that has raged for hundreds of years.

Spacecraft – NASA's Juno orbiter – is currently exploring this giant world.

Did You Know?

There are no rockets powerful enough to hurl a spacecraft into the outer solar system and beyond. In 1962, scientists calculated how to use Jupiter's intense gravity to hurl spacecraft into the farthest regions of the solar system. We've been traveling farther and faster ever since.

Saturn

Image Credit :- NASA

Saturn

Saturn is the sixth planet from the Sun and the second largest planet in our solar system.

Adorned with thousands of beautiful ringlets, Saturn is unique among the planets. It is not the only planet to have rings—made of chunks of ice and rock—but none are as spectacular or as complicated as Saturn's.

Like fellow gas giant Jupiter, Saturn is a massive ball made mostly of hydrogen and helium.

Did You Know?

Twice every 29 and a half years the great planet Saturn appears ringless.

This is an optical illusion:
Earthlings cannot see Saturn's rings when the rings are edge-on as viewed from the Earth. They are barely visible through powerful telescopes.

Uranus

Uranus

The first planet found with the aid of a telescope, Uranus was discovered in 1781 by astronomer William Herschel, although he originally thought it was either a comet or a star.

It was two years later that the object was universally accepted as a new planet, in part because of observations by astronomer Johann Elert Bode. Herschel tried unsuccessfully to name his discovery Georgium Sidus after King George III. Instead the scientific community accepted Bode's suggestion to name it Uranus, the Greek god of the sky, as suggested by Bode.

Interesting Facts About Uranus

- Uranus is known as the "sideways planet" because it rotates on its side.
- Uranus was discovered in 1781 by William Herschel.
- Uranus was the first planet found using a telescope.
- Uranus is an Ice Giant planet and nearly four times larger than Earth.
- Uranus has 27 known moons, most of which are named after literary characters.
- Like Saturn, Jupiter and Neptune, Uranus is a ringed planet.

Did You Know?

Uranus' unique sideways rotation makes for weird seasons. The planet's north pole experiences 21 years of nighttime in winter, 21 years of daytime in summer and 42 years of day and night in the spring and fall.

Neptune

Image Credit :- NASA

Neptune

Dark, cold and whipped by supersonic winds, ice giant Neptune is the eighth and most distant planet in our solar system.

More than 30 times as far from the Sun as Earth, Neptune is the only planet in our solar system not visible to the naked eye and the first predicted by mathematics before its discovery. In 2011 Neptune completed its first 165-year orbit since its discovery in 1846.

NASA's Voyager 2 is the only spacecraft to have visited Neptune up close. It flew past in 1989 on its way out of the solar system.

Did You Know?

Neptune is our solar system's windiest world. Winds whip clouds of frozen methane across the planet at speeds of more than 2,000 km/h (1,200 mph)—close to the top speed of a U.S. Navy F/A-18 Hornet fighter jet. Earth's most powerful winds hit only about 400 km/h (250 mph).

Kuiper Belt

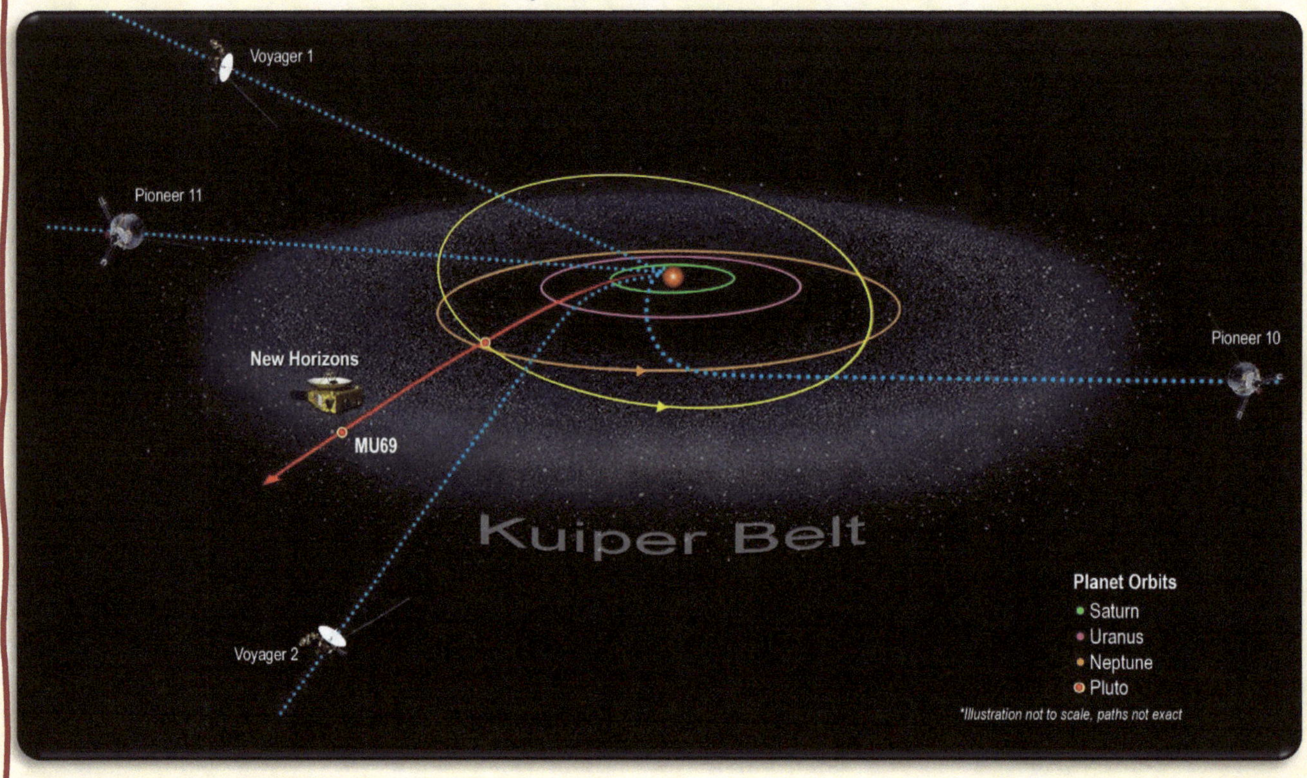

Image Credit :- NASA

Kuiper Belt

Just outside of Neptune's orbit is a ring of icy bodies. We call it the Kuiper Belt.

This is where you'll find dwarf planet Pluto. It's the most famous of the objects floating in the Kuiper Belt, which are also called Kuiper Belt Objects, or KBOs.

There are bits of rock and ice, comets and dwarf planets in the Kuiper Belt. Besides Pluto and a bunch of comets, other interesting Kuiper Belt Objects are Eris, Makemake and Haumea. They are dwarf planets like Pluto.

The Kuiper Belt is a region of leftovers from the solar system's early history. Like asteroid belt, it has also been shaped by a giant planet, although it's more of a thick disk (like a donut) than a thin belt. The Kuiper Belt is truly a frontier in space -- it's a place we're still just beginning to explore and our understanding is still evolving.

Where is the Kuiper Belt?

The inner edge of the Kuiper Belt begins at the orbit of Neptune, at about 30 AU from the Sun. (1 AU, or astronomical unit, is the distance from Earth to the Sun.)

The inner, main region of the Kuiper belt ends to around 50 AU from the Sun. Overlapping the outer edge of the main part of the Kuiper Belt is a second region called the scattered disk, which continues outward to nearly 1,000 AU, with some bodies on orbits that go even farther beyond.

Dwarf Planet – Pluto

Image Credit :- NASA

Dwarf Planet – Pluto

Pluto's classification as a planet has had a history of changes. Since 2006, per the International Astronomical Union's planetary criteria, Pluto isn't considered a planet because it hasn't cleared the neighborhood around its orbit of other objects. However, it does meet IAU's criteria for what constitutes a dwarf planet.

Pluto – which is smaller than Earth's Moon – has a heart-shaped glacier that's the size of Texas and Oklahoma. This fascinating world has blue skies, spinning moons, mountains as high as the Rockies, and it snows – but the snow is red.

Interesting Facts About Pluto

- Pluto is only about 1,400 miles wide. At that small size, Pluto is only about half the width of the United States.
- Pluto is about 3.6 billion miles away from the Sun and has five moons.
- Pluto's atmosphere is thin and composed mostly of nitrogen, methane and carbon monoxide.
- Pluto and its largest moon, Charon, are so similar in size that they orbit each other like a double planet system.
- On average, Pluto's temperature is –387°F (–232°C), making it too cold to sustain life, but it does have a heart-shaped glacier bigger than Texas.

Dwarf Planet - Ceres

Dwarf Planet - Ceres

Dwarf planet Ceres is the largest object in the asteroid belt between Mars and Jupiter and the only dwarf planet located in the inner solar system. It was the first member of the asteroid belt to be discovered when Giuseppe Piazzi spotted it in 1801. And when Dawn arrived in 2015, Ceres became the first dwarf planet to receive a visit from a spacecraft.

Called an asteroid for many years, Ceres is so much bigger and so different from its rocky neighbors that scientists classified it as a dwarf planet in 2006. Even though Ceres comprises 25 percent of the asteroid belt's total mass, tiny Pluto is still 14 times more massive.

Ceres is named for the Roman goddess of corn and harvests. The word cereal comes from the same name.

Did You Know?

Ceres was found by an astronomer searching for a star. He thought he found a comet, but with the help of other astronomers decided it was a planet. As more objects were found between Mars and Jupiter, scientists decided Ceres should be called an asteroid—the largest in the region we now call the main asteroid belt. Then, in 2006, Ceres was reclassified as a dwarf planet—the closest one to Earth.

Dwarf Planet - Makemake

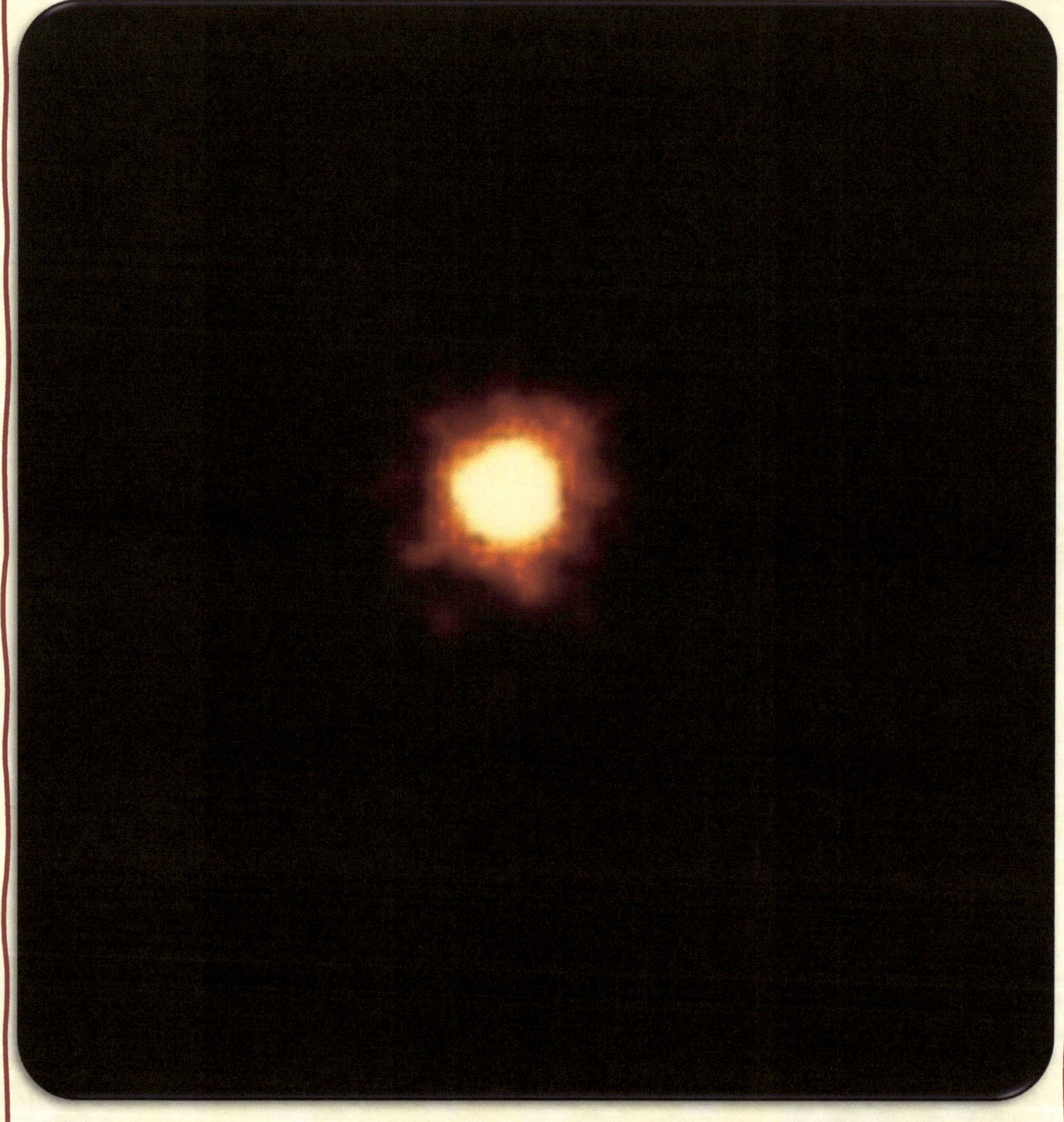

Dwarf Planet - Makemake

Along with fellow dwarf planets Pluto, Eris and Haumea, Makemake is located in the Kuiper Belt, a region outside the orbit of Neptune. Slightly smaller than Pluto, Makemake is the second-brightest object in the Kuiper Belt as seen from Earth (while Pluto is the brightest). It takes about 305 Earth years for this dwarf planet to make one trip around the sun.

Makemake holds an important place in the history of solar system studies because it-along with Eris-was one of the objects whose discovery prompted the International Astronomical Union to reconsider the definition of a planet and to create the new group of dwarf planets. Makemake was named after the Rapanui god of fertility.

Quick Facts	
Day	22.48 hours
Year	305.34 Earth years
Radius	~ 444 miles / 715 kilometers
Planet Type	Dwarf
Confirmed Moons	0
Provisional moons	1

Dwarf Planet - Haumea

Image Credit :- NASA

Dwarf Planet - Haumea

Located in the Kuiper Belt beyond Neptune's orbit, the dwarf planet Haumea is an oval-shaped object with a radius of about 385 miles (just under 10 times smaller than Earth), and two moons, Namaka and Hi'iaka. A day on Haumea lasts only four Earth hours, making it one of the fastest rotating large objects in our solar system.

Originally designated 2003 EL61 (and nicknamed Santa by one discovery team), Haumea resides in the Kuiper belt and is roughly the same size as Pluto. Haumea is one of the fastest rotating large objects in our solar system. Its fast spin distorts Haumea's shape, making this dwarf planet look like a football.

Quick Facts	
Day	4 hours
Year	285 Earth years
Radius	~ 385 miles / 620 kilometers
Planet Type	Dwarf
Confirmed Moons	2

Dwarf Planet - Eris

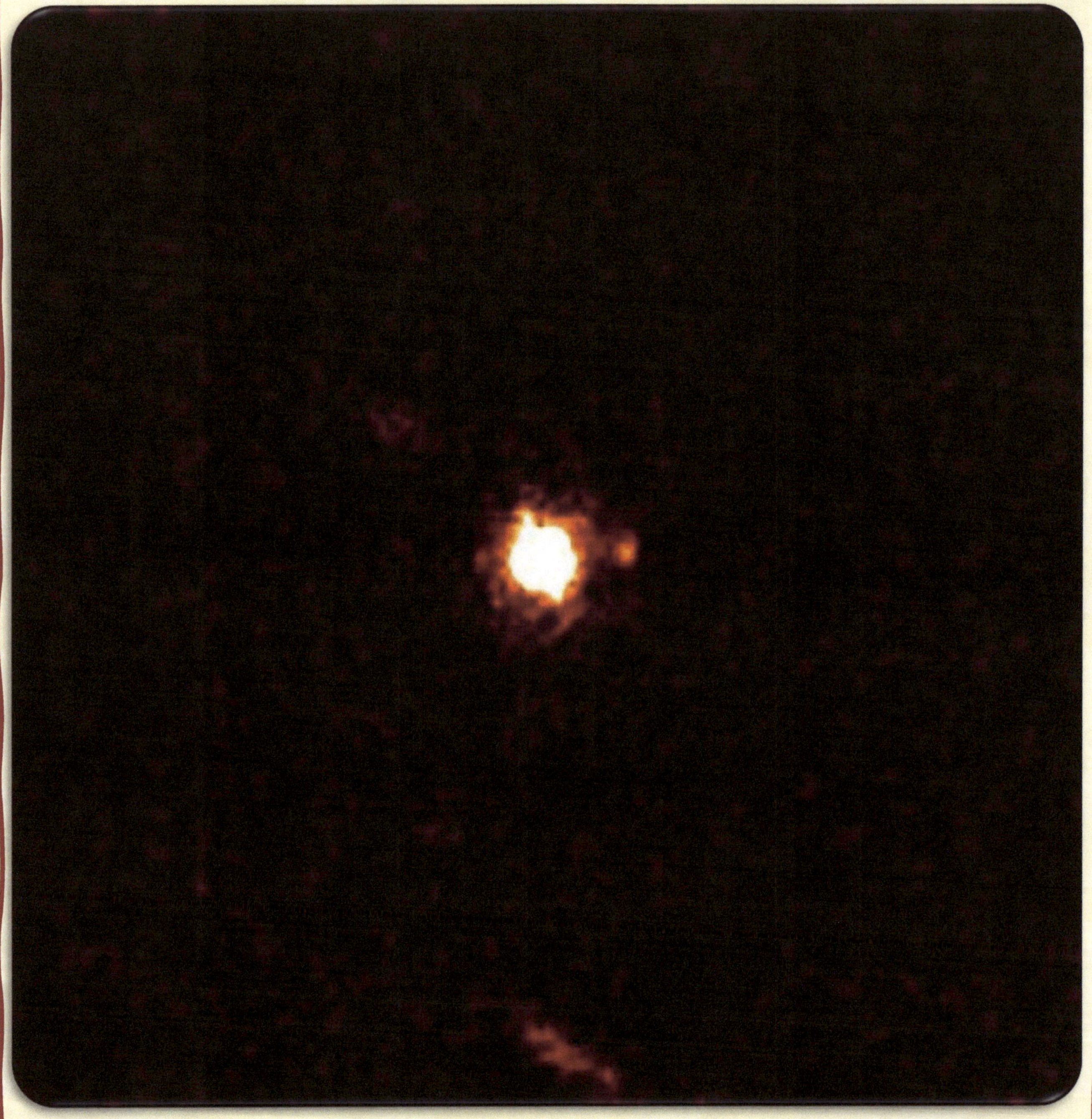

Dwarf Planet - Eris

Eris is one of the largest known dwarf planets in our solar system. It's about the same size as Pluto, but is three times farther from the Sun.

Eris first appeared to be larger than Pluto. This triggered a debate in the scientific community that led to the International Astronomical Union's decision in 2006 to clarify the definition of a planet. Pluto, Eris and other similar objects are now classified as dwarf planets.

Originally designated 2003 UB313 (and nicknamed for the television warrior Xena by its discovery team), Eris is named for the ancient Greek goddess of discord and strife. The name fits since Eris remains at the center of a scientific debate about the definition of a planet.

Quick Facts	
Day	25.9 hours
Year	557 Earth years
Radius	~ 722 miles / 1,163 kilometers
Planet Type	Dwarf
Confirmed Moons	1. Dysnomia

Hypothetical Planet X

Image Credit :- NASA

Hypothetical Planet X

Caltech researchers have found mathematical evidence suggesting there may be a "Planet X" deep in the solar system. This hypothetical Neptune-sized planet orbits our Sun in a highly elongated orbit far beyond Pluto. The object, which the researchers have nicknamed "Planet Nine," could have a mass about 10 times that of Earth and orbit about 20 times farther from the Sun on average than Neptune. It may take between 10,000 and 20,000 Earth years to make one full orbit around the Sun.

The announcement does not mean there is a new planet in our solar system. The existence of this distant world is only theoretical at this point and no direct observation of the object nicknamed "Planet 9" have been made. The mathematical prediction of a planet could explain the unique orbits of some smaller objects in the Kuiper Belt, a distant region of icy debris that extends far beyond the orbit of Neptune. Astronomers are now searching for the predicted planet.

A Roadmap to the Milky Way

Image Credit :- NASA

Beyond Our Solar System

Our Sun is one of at least 100 billion stars in the Milky Way, a spiral galaxy about 100,000 light-years across.

The stars are arranged in a pinwheel pattern with four major arms, and we live in one of them, about two-thirds of the way outward from the center. Most of the stars in our galaxy are thought to host their own families of planets.

The Milky Way galaxy is just one of billion of galaxies in the universe.

The universe is a vast expanse of space which contains all of everything in existence. The universe contains all of the galaxies, stars, and planets. The exact size of the universe is unknown. Scientists believe the universe is still expanding outward.

Other Solar Systems

Our Milky Way Galaxy is just one of billions of galaxies in the universe. Within it, there are at least 100 billion stars, and on average, each star has at least one planet orbiting it. This means there are potentially thousands of planetary systems like our solar system within the galaxy!

Reference :

1. Planets, Moons and Dwarf Planets | NASA
https://www.nasa.gov/content/planets-moons-and-dwarf-planets